Image SEO

Complete Guide

Matthew Lane

Disclaimer

The information provided in *Image SEO: Complete Guide* by Matthew Lane is for general informational and educational purposes only. The author has endeavored to ensure that the content is accurate and up to date at the time of publication. However, the rapidly evolving nature of digital marketing and search engine optimization means that strategies, tools, and practices may change over time. Therefore, the author and publisher make no representation or warranties with respect to the accuracy, applicability, fitness, or completeness of the content of this book.

The author and publisher shall not be held liable for any errors, omissions, or discrepancies contained within this book. Nor shall they be responsible for any losses, injuries, or damages arising from the display or use of this information. The strategies and tips mentioned in this book should be applied according to individual circumstances, and professional advice should be sought where necessary.

This book may contain links to third-party websites or resources. These links are provided for convenience and do not signify endorsement of the content or an affiliation with these sites unless explicitly stated. The author and publisher have no control over, and assume no responsibility for, the content, privacy policies, or practices of any third-party websites or services.

Readers are encouraged to conduct their own research and practice due diligence before implementing any strategies or tools mentioned in this book. The digital marketing field is subject to ongoing changes and advancements, and it is the reader's responsibility to stay informed of the latest trends and updates.

This disclaimer is part of the book. By reading it, you acknowledge that any reliance on the material found in this book is at your own risk.

First paperback edition

ISBN 979-8-9907815-2-8 Paperback

Contents

Introduction

The Crucial Role of Images in the Digital Space

Images are now the digital age's language, changing online communication, sharing, and experiences. Images are crucial for narratives, as they enhance engagement.

The internet has evolved into a visually-driven medium. This evolution parallels the rise of social media platforms and multimedia content. Images are now central to user interaction and content consumption. Visual content dominates the digital landscape. It engages users and enhances online experience.

Research shows that content with relevant images gets 94% more views.[1] Visuals in digital content are vital for capturing attention and boosting message retention. Images can break down complex information, making it more accessible and memorable. Images are crucial in storytelling. They create emotional connections and enhance narrative impact.

MDG, "It's All About the Images [Infographic]," updated 2018, https://www.mdgsolutions.com/learn-about-multi-location-marketing/its-all-about-the-images-infographic/.

In SEO (search engine optimization), using images boosts website visibility and engagement. Images improve website aesthetics, decrease bounce rates, and boost SERP (search engine results page) rankings. However, beyond these functional benefits, images have a profound cognitive impact on users. Images influence perceptions, evoke emotions, and drive actions. They are powerful tools for digital marketers and content creators.

Decoding Image SEO: Beyond the Basics

Image SEO extends beyond mere image visibility in search engine results. It optimizes images to improve SEO and user experience on websites. This section will explore image SEO's importance in digital marketing. There are many facets to image SEO.

Traditionally, search engines could not "see" images the way humans do. They relied on text cues like file names, alt text, and nearby content. They used these cues to understand and index images. Technological advancements, like AI and machine learning, have improved image comprehension in search engines.

Modern image SEO must account for these technological advancements by users optimizing the image files by adjusting format, size, and metadata and by ensuring seamless integration with the content strategy. The integration considers the image's relevance to the text. It also evaluates user experience and thematic enhancement.

Search engines are evolving. They increasingly prioritize user experience. Images play a crucial role in this shift. An optimized image can improve a page. It becomes more appealing to visitors. Search engines recognize and reward this.

We will discuss image SEO mechanics in detail. This includes technical and strategic optimization for web images. This book provides a comprehensive guide to mastering image SEO. It covers image indexing and optimization basics. It also discusses advanced techniques to enhance visibility and engagement. The aim is to help readers navigate the evolving digital landscape.

CHAPTER 1

Fundamentals of Image SEO

Understanding How Search Engines View Images

Understanding Image SEO fundamentals is essential to improving website visibility in search results. This chapter provides an overview of how search engines interpret images. It also covers the key elements of effective image optimization.

Search engines aim to find the best and most useful content for users. While they excel at analyzing text, understanding images presents unique challenges. Search engines don't "see" images like humans. They use text. They use HTML (HyperText Markup Language) elements and image attributes. They use them to understand and index visual content.

Advanced algorithms and machine learning models let search engines analyze an image's visual elements. They can identify objects, scenes, and even emotions. Technology is constantly evolving to improve image interpretation and match images to user queries.

Example Image Recognition Technology:

- Google Lens: Google Lens is an example of image recognition technology that allows users to search for anything they can see. For instance, if you point your smartphone camera at a landmark, Google Lens can provide historical information about it. Similarly, if you take a photo of a restaurant's sign, Google Lens can pull up reviews, hours, and other details. Another use is in online shopping. Users can take a photo of an item they like. Google Lens will then search for the item to buy online or suggest similar products.

- Facial Recognition Systems: Technologies used by platforms like Facebook identify and tag people in photos. When you upload a picture, Facebook's technology can suggest tagging friends who appear in the image. It does this based on its recognition algorithms. These algorithms compare the faces in the photo with previously tagged images.

- Image Search Engines: Services like TinEye offer reverse image searches. When you upload an image to TinEye, it uses image recognition to find where the image has appeared online. It provides information about the image's source, other available sizes, and any modifications.

- Visual AI for E-commerce: Amazon Rekognition is a service that makes it easy to add image analyses to applications. E-commerce sites use it to search and filter products through images. Rekognition can find similar items in a catalog by analyzing an image uploaded by a user.

These technologies rely on advanced neural networks. They have been trained on vast image datasets. This training allows them to see and learn from patterns, much like the human brain. This capability is changing how we interact with digital content. It is key in fields such as digital marketing, security, and online retail.

Alt Text, File Names, and Captions

A search engine crawls a web page to find images and their metadata. This helps it understand and catalog the content. Metadata like alt text, captions, and file names gives search engines context about the image. This helps them understand its content and relevance to the text.

- Alt text (alternative text) is a critical component, serving as a descriptor of the image content for people who cannot see the image. So you should describe it in a way so that they can visualize it. It also helps search engines index images accurately. Alt text should try to describe the image as accurately as possible. You can also include the primary or secondary keywords in the alt text to improve SEO. Be careful not to force it, and always avoid keyword stuffing. When in doubt, make it for humans, and make it sound natural.

- File names for images are crucial for SEO, as they help search engines understand and rank the content effectively. It's important to use descriptive, keyword-rich file names that accurately reflect the image content. Instead of generic names like "IMG_1234.jpg," opt for clear and specific descriptors such as "homemade-chocolate-chip-cookies.jpg." This approach improves SEO. It also helps users by making file contents predictable. Use hyphens to split words, search engines read them better than underscores or spaces. Using short file names helps, and they should be relevant and easy to read. This change can greatly boost your image's search visibility.

- Captions for images play a vital role in both enhancing user engagement and improving SEO. Alt text describes visual content for search engines and accessibility tools, but captions add extra context, draw attention, and can engage users directly.

Example Alt Text, File Name, and Caption

I want to show an example image and potential alt text, file name, and caption. These are critical aspects of image SEO, and I want this to help your understanding. Also, as a neat treat, this is an image generated with DALL-E by ChatGPT (mentioned later in this book).

Example Alt Text

Article/image topic: Annual Coffee Consumption Trends

Alt Text: “Infographic showing global coffee consumption trends in 2023, including top coffee-drinking countries, average consumption per capita, and popular coffee types.”

Why This Alt Text Works:

- Descriptive: It clearly explains what the image is about, which is essential for search engine indexing and beneficial for visually impaired users.

- Informative: It includes key details that someone who cannot see the infographic might need to understand its content.

- SEO-friendly: It includes keywords that someone might search for, such as “coffee consumption trends 2023.”

Example File Name

File Name: 2023-coffee-consumption-trends-infographic.jpg

Why This File Name Works:

- Clear and Descriptive: It tells you exactly what the image is about just from the file name, which is useful for organization and SEO.

- Hyphenated: Using hyphens to separate words in the file name makes it easier for search engines to read and understand, improving SEO.

- Keywords Included: It includes keywords that are relevant to the image, which can help improve the image's visibility in search results.

Both elements (alt text and files names) are crucial for SEO and accessibility, making your content more discoverable and usable for all audiences.

Example Caption

"Explore the world of coffee with our detailed infographic. It shows global consumption patterns. Discover which countries lead in caffeine indulgence. Also see the variety of coffee preferences across continents. These reflect a universal passion for this beloved beverage."

Why This Caption Works:

- Descriptive and Informative: The caption provides a snapshot of what the infographic represents, guiding the viewer to understand the data's focus on coffee consumption around the world.

- Engages the Audience: By inviting viewers to "explore" and "discover," the caption encourages interaction and further exploration of the details within the infographic.

- Relevant Keywords: The use of terms like "global consumption patterns," "caffeine indulgence," and "coffee preferences" ties in relevant keywords that could be beneficial for SEO without resorting to keyword stuffing.

How to Edit Image Alt Text, File Names, and Captions

Editing Image Alt Text

The alt text should say what the image shows. It should use keywords naturally. Here's how to edit it:

- For Websites:
    - HTML: If you're coding directly in HTML, you can add or edit the alt attribute in the image tag, like so: <img src="image.jpg" alt="Your descriptive text here">. When you're setting up or editing the alt text in HTML, the alt attribute is inside the <img> tag. You insert your descriptive text there. If you're updating an existing image, replace the old alt text with your new text. For example, if the existing tag

is <img src="old-image.jpg" alt="Old description">, and you want to update it, you would change it to something more descriptive like <img src="old-image.jpg" alt="A clear sunset over the mountains">. This makes it clear that you are updating the text in the alt attribute. You do this to enhance the image description for better accessibility and SEO.

  - Content Management Systems (CMS): In platforms like WordPress, when you upload or select an image in the Media Library, you'll typically see a field for alt text where you can enter or edit the text.

- For E-commerce Platforms:
  - Platforms like Shopify or WooCommerce have specific fields for alt text when you upload or manage product images. Look for an "Alt Text" or "Image Description" field.

Editing File Names

File names should be descriptive and include keywords that reflect the image subject.

- Before Upload: The simplest way to edit a file name is before uploading it to the web. Right-click the file on your computer, select "Rename," and then type the new name.

- After Upload: Some CMS platforms allow you to change the file name after upload, but this is less common because it can affect URLs and potentially break existing links to the image. It's best to set the file name correctly before uploading.

Editing Captions

Captions provide context to the images and can enhance user engagement and SEO with descriptive and keyword-rich text.

- In HTML: You can add a caption directly under an image with a <figcaption> tag inside a <figure> element.
- In CMS Platforms: When you insert an image, there's often an option to add a caption directly in the image settings. Many blogging platforms like Medium or article publishing on LinkedIn also allow you to easily insert a caption.
- For Social Media: When posting an image, use the post's text area to add your caption before publishing.

Best Practices for Editing:

- Be as descriptive and specific as possible without being overly verbose.
- Use relevant keywords naturally, as if you're writing a sentence.
- Consider users who may be using screen readers. The alt text should convey the same function and purpose as the image.
- Don't repeat the same information in the alt text, file name, and caption. Each should offer unique value.

Remember, for SEO, each image on your site is a chance to boost your site's visibility and accessibility. Crafting alt text, file names, and captions well is not just about following SEO best practices. It also makes the web more inclusive and engaging for all users.

Note: Keyword stuffing in image SEO happens when too many keywords are crammed into the image tags. This includes the alt text, file names, and even the surrounding descriptive text or captions. It's all an attempt to manipulate a website's search engine rankings. Search engines see this practice as spammy. It can lead to penalties or lower rankings. They think it harms user experience and can mislead search engine indexing.

Example of Keyword Stuffing in Image SEO

Imagine you have a website selling organic coffee. Here's how keyword stuffing might appear in your image SEO:

- File Name: organic-coffee-buy-organic-coffee-best-organic-coffee-coffee-shop-online.jpg
- Alt Text: Buy organic coffee, best organic coffee, cheap organic coffee, organic coffee online, organic coffee near me
- Image Caption: Get the best organic coffee here! Buy organic coffee, organic coffee sale, top organic coffee flavors, organic coffee benefits, why organic coffee

Why This Is Problematic:

- Repeating similar or identical phrases ("organic coffee") does not add additional information or value.
- For a visually impaired person using a screen reader, hearing this overstuffed and repetitive alt text would be confusing and not helpful.
- Search engines, like Google, are sophisticated enough to recognize keyword stuffing and may penalize your site by ranking it lower in search results.

Good image SEO should use concise, descriptive language. It should accurately describe the image and include relevant keywords naturally. This approach improves both user experience and search engine visibility.

The Anatomy of an Optimized Image

An optimized image has a blend of quality, relevance, and technical precision. It enhances user experience and search engine rankings.

- File Size and Resolution: Large images can slow down web pages. They hurt user experience and SEO. Optimizing the file size without compromising quality is essential. Tailor the resolution to match the platform's display requirements. This ensures clear images that are not overly large.

- File Format: The choice of image format (JPEG, PNG, WebP, etc.) affects both the image's appearance and its file size. Select the correct format for optimizing images by considering content and quality.

- Compression Techniques: Effective image compression reduces file size while maintaining visual quality. Compression tools and algorithms keep images looking good. They also make them load fast by shrinking the file size. This is an easy thing to do, and everyone should try to do it. There are many free tools available online. It's often as easy as uploading the image, converting the file, then downloading the file. Remember also to be sure the images are clear. You do not want to reduce the file size of the image to the point where it is blurry.

- Metadata Optimization: Beyond alt text, images can include metadata like EXIF data (mentioned later in this book, camera settings, geolocation, etc.), which can be stripped to reduce file size if not relevant. Using metadata improves SEO and organizes image assets.

CHAPTER II

Technical Aspects of Image Optimization

In the digital world, speed and efficiency are crucial. Optimizing images is not just best practice; it's necessary. This chapter discusses image SEO details. It focuses on decisions that affect website speed and search engine rankings.

Choosing the Right Format: JPEG, PNG, WebP, and Beyond

An image's file format determines how its data is stored and displayed. Choosing the correct format is crucial for maximizing web image quality and performance.

- JPEG, short for Joint Photographic Experts Group, is mainly used for photos and realistic paintings. It is a go-to format for its efficient compression. Reducing file size significantly speeds up website load times. However, this has a cost: JPEG uses lossy compression, which means some data will be lost. It sacrifices some image quality for smaller file sizes. Detailed precision is less important.

- PNG (Portable Network Graphics) is favored for its lossless compression. It ensures that image quality is not compromised after saving and compression. PNGs support transparency. They are suitable for logos, icons, and graphic designs needing sharp edges and clear distinctions. The trade-off with PNGs is larger file sizes compared to JPEGs, which can affect loading times if not used judiciously. For nontransparent images, it is generally recommended to convert the file to JPEG if using on a web page.

- WebP is a modern image format made by Google. It has better compression, so it is a great choice for both lossy and lossless compression. It supports transparency like PNG and can achieve smaller file sizes at comparable quality levels to JPEG. This makes WebP a versatile choice for various web images, offering a good balance between quality and performance.

- Emerging Formats: New formats like AVIF (AV1 Image File Format) are entering the market, offering even better compression and quality characteristics. Staying updated on these developments is crucial. These new formats can provide big benefits for image SEO, like speed and visual quality.

Mastering Image Compression Techniques

Image compression reduces the file size of images, making them faster to load on web pages. It's a critical step in image optimization. It affects both user experience and search engine ranking.

- Lossy vs. Lossless Compression: Lossy compression reduces file size by deleting some image data. This deletion can affect image quality. It's useful for large photos. They can tolerate slight quality loss for big size cuts. Lossless compression reduces file size without losing quality. It is ideal for detailed graphics or when image quality must remain pristine.

- Advanced Compression Tools: Tools like Compressor.io, TinyPNG, and Adobe Photoshop offer advanced compression capabilities. Compressor.io efficiently reduces image file size while maintaining quality. It is an essential tool for SEO.

- Automated Image Optimization: Services like Cloudinary and Imgix automate optimization. They adjust image size, format, and compression based on the end user's device. Images are delivered quickly. This speeds up load times and improves user experience on many platforms and devices.

Understanding and applying image optimization techniques can boost a website's SEO. This can significantly improve its performance. Webmasters optimize site images by selecting the correct format. They master compression techniques. This leads to faster load times, better user engagement, and higher search rankings.

Advanced Image SEO Strategies

CHAPTER

AI and visual search improve image SEO. They do this by boosting performance and visibility in the digital world.

AI and Image Optimization: The Next Frontier

AI revolutionizes image SEO by creating and optimizing images for better search performance.

- AI can now create images that perform well in search rankings. AI-generated images can be adjusted for SEO to meet size, format, and relevance. The images are typically completely original and can be created very quickly, saving time and money.

- AI can make descriptive, keyword-rich file names and alt text. These are vital for SEO success. AI tools can analyze an image's content. Then they can add metadata. This metadata describes the image and also includes SEO keywords. This can boost the image's visibility in search results.

- Enhanced image tagging and categorization uses AI for automated tagging and categorization, which helps index images, making them easier for search engines to find. AI systems analyze images and assign detailed tags and categories. This enriches the metadata and improves searches.

AI-Enhanced Image Creation Tools

- DALL-E is an AI program by OpenAI. It can generate unique images from text. DALL-E can make high-quality SEO-friendly images for specific content needs. It integrates with GPT-4. This allows for advanced image creation. The images can enhance content and SEO.

- Adobe Photoshop now has AI-driven tools. They can automate and optimize image creation. This makes it easier to make images that are both pretty and good for SEO.

- Canva is a design tool that uses AI to make web images. It gives tips on design elements that boost engagement. It also helps with SEO best practices.

AI-Driven SEO Optimization Tools

- Google Cloud Vision API is a machine learning tool. It analyzes images and pulls useful data for SEO. This data includes object recognition, which can make alt text and images easier to find.

- seoClarity uses AI to analyze images. The tool gives insights and tips for optimizing image content. This helps improve ranking in image search results.

- GumGum is an AI-powered visual analysis platform. It offers services that improve image engagement and performance. These services align with SEO goals and improve user experience.

Visual Search Optimization Tools

- Google Lens is Google's visual search tool. It helps users search the web with images. It provides insights into how search engines interpret images and how to optimize them for better search performance.

- Pinterest Lens is a tool for visual discovery on Pinterest. It helps SEO practitioners understand how to optimize images for image-centric platforms.

- Bing's Visual Search offers insights, showing how Visual Search processes and matches images. This helps optimize images for better discoverability and ranking.

Comprehensive Image SEO Platforms

- Cloudinary is a full image management solution. It offers AI-powered features for automatic image tagging, optimization, and delivery. These features ensure that images are served in the best format and size on different devices and platforms.

- Imgix provides real-time image processing and optimization. It delivers images optimized for the viewer's device. This enhances load times and the user experience and helps SEO.

Anyone can use these tools to make AI-generated images. They can then optimize the images for SEO. This will improve their website's performance in visual search engines. This approach improves a website's look. It also boosts its performance in search rankings.

Businesses can use the strategies in this chapter, leveraging AI to create and improve images. The goal is to make the images look good and fit SEO objectives. Also, optimizing for visual search can unlock new opportunities. It can improve online visibility and engagement. This makes image SEO a powerful tool in digital marketing.

CHAPTER IV

SEO for Different Types of Visual Content

This chapter focuses on improving many visual content types. These include infographics, memes, and animated content. Each visual content type has unique features. To boost online visibility and engagement, specific SEO strategies are needed for each type.

Optimizing Infographics, Memes, and Animated Content

Infographics, memes, and animated content are powerful tools. They convey information and engage users. However, their SEO potential is often underutilized.

- Infographics: These combine data and design to simplify complex information. For SEO, make sure infographics are shareable. Also, they need embedded codes that link back to the original site. The alt text and file names should describe the infographic's content. They should not just describe its appearance.

- Memes: Though primarily used for entertainment, memes can boost engagement. They should be relevant to the content. They should include trending or searchable phrases as text within the image. Also, they should have alt text for context.

- Animated Content (GIFs, short videos): These are effective for grabbing attention. They should be optimized with descriptive file names and alt text. If hosting on your own site, compress them for quick loading. Integrate them with the rest of the content to enhance the user experience.

Video SEO: An Integral Part of Image Optimization

Videos are now dominant in digital marketing. Optimizing them is crucial for improving search visibility and user engagement.

- Thumbnails: These are the first things viewers see. An optimized thumbnail should be compelling, high-quality, and relevant to the video content. It acts as a preview, influencing the click-through rate.

- Transcripts and Captions: Providing text transcripts and captions for videos enhances accessibility and indexability. They let search engines crawl the video's content. This improves its chances of ranking for relevant queries.

- Video Metadata: Like with images, videos need metadata, which should have many keywords and be descriptive. This information helps search engines understand and index the content appropriately.

Example Video Metadata

Title: The Art of Latte: Barista Techniques and Tips

Description: "Dive into the craft of latte art with our expert baristas as they reveal secrets to perfecting your pour. This video covers everything from frothing milk to complex design techniques. They can turn your morning coffee into a canvas."

Tags: latte art, barista skills, coffee making, milk texturing, coffee designs, espresso techniques, latte tutorial

Category: How-to & Style

Thumbnail URL: https://example.com/thumbnails/latte-art.jpg

Video URL: https://example.com/videos/the-art-of-latte.mp4

Duration: 5:30

Upload Date: 2024-04-14

Language: English

Subtitles/CC: Available in English and Spanish

Location: New York, NY, USA (if relevant for local SEO)

Content Rating: G

View Count: 150,000 (if the platform supports metadata updates after upload)

Likes/Dislikes: 4,000 likes, 100 dislikes (if applicable)

Embedding Options: Allowed

License: Standard YouTube License

Video File Metadata (not always visible to viewers):

- Codec: H.264
- Frame Rate: 30fps
- Resolution: 1920x1080

- Aspect Ratio: 16:9

Comments: Enabled

This metadata helps search engines understand the video's topic. It helps with indexing and ranking the video for relevant searches. Better metadata is more likely to help the video be found. It will be found by users searching for related content.

Editing Video Metadata

Editing video metadata usually means changing or adding info to the file properties. This process can vary by platform, such as YouTube, Vimeo, or a desktop video file. But here are general steps and tips for editing video metadata:

For Video Files on Your Computer:

1. In Windows, you can right-click the file and select "Properties," then go to the "Details" tab. On a Mac, right-click and choose "Get Info."

2. Input or update information in fields such as title, subtitle, comments, tags, and more. Some fields may not be editable depending on the file format.

3. Once you've updated the necessary fields, make sure to apply the changes and save them.

For Videos on YouTube:

1. Log in to your YouTube Studio.
2. Click on the video you wish to edit.
3. Go to the "Details" section where you can update the title, description, tags, and category.
4. Add cards and end screens, which are interactive elements that can direct viewers to other videos, playlists, websites, or calls to action.
5. Go to Advanced Settings. Here, you can adjust settings such as comments, licensing, language, and more.
6. Don't forget to save your changes after editing.

For Videos on Other Platforms:

1. Look for an option to edit or manage your video after it's been uploaded.
2. Fill in or revise the metadata fields available, which might include title, description, tags, categories, and privacy settings.
3. Make sure your changes are saved or published.

Tools for Editing Metadata:

- Video Editing Software: Some advanced video editing software, like Adobe Premiere Pro, allows you to edit metadata before exporting the final video file.
- Dedicated Metadata Editors: There are tools specifically designed for editing metadata, such as MetaX, MP3Tag (which also works for video files), and others.

Tips for Editing Metadata:

- Consistency: Ensure that your metadata is consistent across platforms to help with recognition and SEO.
- Keywords: Incorporate relevant keywords in your title, tags, and descriptions for better search engine visibility.
- Clarity: Make your metadata clear and descriptive to improve user understanding and engagement.

Always remember: metadata is a powerful tool. It's not just for organization but also for SEO and engaging viewers. It's worth spending the time to ensure it is as informative and accurate as possible.

Use these focused SEO strategies. They can optimize many kinds of visual content. This will improve your content's search rankings and user engagement. This approach boosts the visibility of the visual content. It also helps the website's SEO.

CHAPTER

Unconventional Tactics and Hidden Gems in Image SEO

This chapter explores the deep, often overlooked parts of image SEO. It covers new strategies and lesser-known tips to give your content an edge in a tough digital space. It aims to give you the tools and knowledge to push your image SEO beyond the usual limits by using unconventional tactics and by finding the hidden gems in image attributes and formats. These strategies show the importance of innovation and attention to detail. They are key for crafting an SEO approach that stands out in today's visually-driven online world.

Leveraging SVGs for Scalable SEO

Scalable Vector Graphics (SVGs) are not just for crisp visuals at any screen resolution; they also offer SEO benefits that other file types cannot match.

- SEO Benefits of SVGs: SVG files are XML-based, which means they can be indexed and understood by search engines much like HTML. For example, if you embed an SVG image of a pie chart in your web page's code, you can include titles and descriptions within the SVG. Search engines can read them.

- Vector Graphics for Responsiveness: SVGs scale without loss of quality, which makes them perfect for a responsive design. A well-implemented SVG for a company logo ensures that your branding looks sharp on all devices. This includes watches and widescreen monitors. And it does so without needing multiple file versions.

Hidden SEO Power in Image Metadata

Beyond the visible image data, metadata can provide a wealth of SEO opportunities that many neglect.

- EXIF Data for Local SEO: EXIF data can include GPS coordinates, which can be leveraged for local SEO. A real estate website might use EXIF data to highlight properties' locations, reinforcing geographic relevance in search results.

- IPTC for Comprehensive Detailing: IPTC metadata can contain a wealth of information such as headlines, descriptions, and copyright notices. An online art gallery could utilize IPTC to include artist names, artwork titles, and detailed descriptions, enriching the image data for search engines.

The Untapped Potential of Progressive Images

Progressive images offer a user experience advantage. They can help SEO by improving perceptions of page load speed. Instead of loading line-by-line, progressive JPEGs load in waves of increasing detail, making a page feel faster. An e-commerce site shows a product gallery, so it can use progressive JPEGs. This makes images appear to load instantly, even if the full detail isn't clear yet.

Let's discuss the finer points of editing for image SEO, including SVGs, metadata, and progressive JPEGs. It can be somewhat technical. Here's a guide on how to make these edits:

Editing SVGs for SEO

SVG files are essentially text files that describe an image in XML format. To edit them

1. Open the SVG File: Use a text editor like Notepad++ or an IDE like Visual Studio Code.

2. Add Titles and Descriptions: Insert <title> and <desc> tags within the SVG tags to add SEO-friendly titles and descriptions.

    a. Example:

        <svg> <title>Interactive Map of California</title> <desc>Map highlighting the major tourist attractions in California.</desc> ... </svg>

3. Include Keywords: Ensure the titles and descriptions contain relevant keywords for your content.

Editing Image Metadata

EXIF and IPTC metadata can be edited using photo editing software or specialized tools.

1. Using Photo Editing Software: Programs like Adobe Photoshop or Lightroom allow you to view and edit metadata under the "File Info" or "Metadata" panels.

2. Dedicated Metadata Editors: Tools like ExifTool or MetaX can edit this data. They have fields where you can enter information like location, copyright, creator details, etc.

Creating Progressive JPEGs

Progressive JPEGs are created during the saving process of the image.

1. Adobe Photoshop: When saving a JPEG, select the option "Progressive" from the JPEG options dialog box.

2. Online Tools: Websites like TinyJPG can convert standard JPEGs to progressive JPEGs.

Steps for Each Element

- SVG: Edit directly in the code, or use an SVG editor.

- Metadata: Use software that allows metadata editing, fill in relevant fields, and save.

- Thumbnails: Resize, optimize, and save using a "Save for Web" feature for quick loading.

- Progressive JPEGs: Choose the progressive option when saving your JPEGs for a smoother loading experience.

Remember to back up your original images before making changes. Also, test changes to ensure they show correctly and improve the user experience. These details of image SEO can greatly impact your site's performance and rankings. But only if done well.

CHAPTER
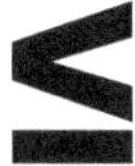

Image SEO Tools and Technologies

In image SEO, the right tools and tech can greatly speed up optimization. They also improve image quality and search performance. Chapter 6 is about the key tools and tech that help image SEO.

Comprehensive Image SEO Platforms

- Cloudinary is a platform for image management. It provides solutions for uploading, storing, managing, and optimizing images. Cloudinary uses advanced algorithms. It applies the best format, compression, and resolution. This improves loading times and SEO.
- Imgix is similar to Cloudinary. It is a powerful image processing service that optimizes images in real time.

It adjusts images based on the viewing context. This ensures they display well on all devices and platforms. This is crucial for strong SEO.

AI-Powered Image Optimization Tools

- DALL-E is by OpenAI. It is an AI that generates images. It can create detailed and relevant images from text. For SEO, DALL-E can make unique, good images. They can be customized to have SEO-friendly attributes. These include relevant file names and alt text. This is my favorite AI image-generation tool. You need ChatGPT-4 to use this application.

- Let's Enhance uses AI to upscale images without losing quality. This is useful for improving website images. It enhances user experience and SEO.

Image Compression and Editing Tools

- Compressor.io is known for its strong compression. It reduces image file size without reducing quality. This is vital for speeding up websites and improving SEO. This is my favorite image compression tool. It's free and very quick and easy to use.

- Adobe Photoshop offers many features for editing and optimizing images. It includes options to adjust size, resolution, and format. These are essential for making SEO-friendly images.

SEO-Specific Image Tools

- SEO Clarity provides image SEO insights, suggesting improvements for better search engine visibility. It can analyze an image's performance and recommend changes to enhance its SEO value.

- Screaming Frog SEO Spider is primarily a website crawling tool. But it can also audit image SEO. It does this by analyzing aspects like alt text, file size, and file format. It checks these aspects across a website to ensure that all images are optimized for search engines.

Using these tools well can greatly improve image SEO. They allow for better optimization, management, and performance of web images.

Ethical Considerations and Legal Aspects of Image SEO

CHAPTER

Professionals must navigate the ethical and legal image SEO landscape. Doing so is key for those aiming to optimize their online presence responsibly. This chapter covers the main legal rules. It also covers the ethical dilemmas and best practices for using images in SEO.

Navigating Copyrights and Image Licensing

You need to understand copyright laws. They apply to images used online. This understanding is key to avoiding legal issues and ensuring fair use.

- Copyright Law: Copyrights protect the creative work of individuals, granting the creators exclusive rights to use, distribute, and reproduce the work. For instance, using a copyrighted image without permission on a website can lead to infringement claims. It's crucial to understand the differences between copyrighted material, public domain, and fair use. The public domain has works not protected by copyright that can be used freely. Fair use is a legal doctrine. It allows limited use of copyrighted material without permission. It's typically for purposes like criticism, education, or parody.

- Image Licensing: Different licensing agreements dictate how images can be used. For example, royalty-free images can be used without paying royalties. But they may have usage restrictions on type or volume. Rights-managed licenses offer specific usage rights. They are based on time, location, and manner of use. Creative Commons licenses allow free use of images under certain conditions. Each license type dictates how an image can be used in SEO campaigns.

Ethical Use of Images

Using images ethically goes beyond following the law. It involves the integrity and moral duty of content creators.

- Image Sourcing: Ethical sourcing involves using images that are legally and morally right to use. It respects the rights of the image creators and ensures images are used in context. For example, using an image that shows sensitive subjects, like people in distress, requires great care. Privacy, consent, and potential harm need to be considered.

- Representation and Diversity: Ethically, images should represent diversity and avoid perpetuating stereotypes. This means picking images that show many different people, backgrounds, and scenarios. They should be shown respectfully and inclusively. For instance, if a website is about global cultures, it should use images that show the diversity of those cultures. The images should be accurate and respectful.

Ethical Use of AI in Image Creation and Optimization

AI technology plays a growing role in creating and improving images. This makes ethics more important.

- Transparency in AI-Generated Images: There is a need for transparency when using AI-generated images. Users should be informed if images are AI-generated to avoid deception. For example, a disclaimer might be needed when publishing AI-generated images to clarify their origin in articles or ads.

- Bias in AI Algorithms: AI systems can inadvertently perpetuate biases present in their training data. For image SEO, this means watching how AI-generated images might show or boost stereotypes or inaccuracies. Efforts should be made to use AI tools that are trained on diverse data sets to minimize these biases.

It is crucial to navigate the complex legal and ethical world of image SEO carefully and honestly by understanding and fixing these aspects. SEO practitioners can ensure their practices boost search rankings and uphold ethical standards and legal requirements. This fosters trust and credibility in the digital landscape.

CHAPTER

Future Trends and Innovations in Image SEO

Chapter 8 explores the expected trends and innovations that will shape the landscape of image SEO. This chapter provides foresight, showing how new technologies and user behaviors will shape image optimization. It will help ensure that SEO practitioners are ready for upcoming changes.

The Rise of AI and Machine Learning

AI and machine learning are set to play even more significant roles in image SEO. These technologies will continue to advance, offering the following:

- Automated Optimization: AI algorithms will get better at optimizing images for SEO. They will adjust attributes like size, format, and alt text to match search engine preferences.

- Predictive SEO: Machine learning models could predict future SEO trends. This would allow for proactive image optimization to meet expected search algorithm updates.

Enhanced Visual Search Capabilities

As visual search technologies become more sophisticated, their impact on SEO will grow. Future trends may include the following:

- Image Search as the Main Query Method: AI and machine learning have improved. Users may rely on visual search more and more instead of text. This makes image optimization crucial.

- Integration of Visual and Voice Search: The merging of visual and voice search could lead to new SEO strategies. Images will need to be optimized for both looks and relevance to spoken queries.

New Image Formats and Compression Techniques

Image formats and compression will keep evolving. They are driven by the need for faster, more efficient web experiences.

- Next-generation image formats (formats like AVIF and others still in development) promise better compression and quality. They could revolutionize image SEO by making high-quality images more accessible and faster to load.

- AI-driven compression may offer smarter compression techniques, making adjustments based on content type and user context. This will enhance load times and user experience without sacrificing image quality.

The Impact of Augmented Reality (AR) and Virtual Reality (VR)

AR and VR are starting to affect web content. They are changing how images are used and made better.

- As AR and VR become more popular, images must be made better for these experiences. New SEO techniques for 3D and interactive content may be needed.

- For e-commerce, AR and VR can make product images interactive. This requires optimization for these formats to boost engagement and conversion.

I've aimed to prepare readers for the future of image SEO, highlighting the need to stay ahead of new technology and changing user habits. By knowing and expecting these trends, SEO practitioners can keep their image optimization strategies effective. They can keep them forward-thinking in the ever-changing digital landscape.

Conclusion

Let's reflect on the journey through the complex world of image SEO, capturing the key insights and strategies from the book and consolidating the knowledge. Then I'll offer guidance on how to maintain effective image SEO over time.

Key Takeaways

- Images are critical. They enhance user experience and search engine optimization.
- Technical optimization is about understanding why file size, format, and metadata must be optimized: to improve website performance and search rankings.
- Embracing advanced techniques, including AI-driven optimization and visual search, and adapting to new image formats are crucial in staying ahead in the SEO game.
- Ethical and legal considerations are important. They are key for using images online.

Continuing Your Image SEO Journey

SEO is always changing. This is especially true for image SEO. It needs continuous learning and adaptation to new trends, technologies, and best practices.

- Stay updated with industry changes. Regularly follow SEO and digital marketing blogs. Attend webinars and conferences. Also, join in community discussions to stay informed about the latest trends and updates in image SEO.

- Experiment and innovate. Don't hesitate to try new tools. Experiment with emerging tech. Test different strategies to find what works best for your needs and goals.

- Track image performance using analytics and SEO tools. Use them to monitor how your images appear in search results. Regular analysis helps in identifying opportunities for improvement and adjusting strategies accordingly.

Image SEO Quiz

1. **What is the primary purpose of image SEO?**

   A. To make images look better on a website
   B. To optimize images to improve website load times
   C. To enhance the visibility and ranking of images in search engine results
   D. To increase the size of images on a website

2. **Which image format is known for its efficient compression and quality balance, suitable for web use?**

   A. JPEG
   B. TIFF
   C. BMP
   D. WebP

3. **How do search engines primarily understand the content of an image?**

   A. By analyzing the color scheme of the image
   B. Through the metadata associated with the image, such as alt text and file names

C. By measuring the dimensions of the image
D. By the number of likes or shares the image has on social media

4. **What role does AI play in advanced image SEO strategies?**

A. AI generates high-resolution images only.
B. AI assists in automating image tagging and optimizing metadata.
C. AI reduces the need for images in websites.
D. AI exclusively enhances the aesthetic appeal of images.

5. **Which of the following is NOT a common pitfall in image optimization**?

A. Using overly large image files
B. Neglecting the alt text for images
C. Regularly updating image file formats
D. Over-optimizing images to the point of quality degradation

6. **What is the significance of alt text in image SEO?**

A. It provides a textual alternative to the image for search engines and visually impaired users.
B. It increases the physical size of the image file.

C. It changes the color of the image to be more SEO-friendly.
D. It is only used for decorative purposes and has no SEO value.

7. **In the context of image SEO, why is it important to consider the ethical use of images?**

A. Ethical considerations are only important for legal compliance.
B. Using images ethically can enhance a brand's image and trustworthiness.
C. Ethical use of images speeds up the website's performance.
D. There are no real consequences for unethical use of images.

8. **How does visual search impact image SEO?**

A. Visual search eliminates the need for image SEO.
B. Visual search technologies rely on text-based queries only.
C. Visual search enhances the need for optimized, high-quality, and contextually relevant images.
D. Visual search is unrelated to image SEO and does not impact it.

9. **What future trend is anticipated to have a significant impact on image SEO**?

   A. The discontinuation of all image formats except JPEG
   B. The decline of visual content on the web
   C. The rise of AI and machine learning in automated image optimization
   D. The removal of all images from search engine indexes

10. **Why is it crucial to stay updated with the latest image SEO practices and technologies?**

   A. Image SEO practices and technologies remain constant over time.
   B. Outdated practices are more effective than modern ones.
   C. Search engines frequently update their algorithms, impacting how images are indexed and ranked.
   D. Modern technologies are less reliable for image SEO.

Answers:

1. C - To enhance the visibility and ranking of images in search engine results
2. D - WebP
3. B - Through the metadata associated with the image, such as alt text and file names
4. B - AI assists in automating image tagging and optimizing metadata.
5. C - Regularly updating image file formats
6. A - It provides a textual alternative to the image for search engines and visually impaired users.
7. B - Using images ethically can enhance a brand's image and trustworthiness.
8. C - Visual search enhances the need for optimized, high-quality, and contextually relevant images.
9. C - The rise of AI and machine learning in automated image optimization
10. C - Search engines frequently update their algorithms, impacting how images are indexed and ranked.

Appendix

Resource Directory

This Resource Directory has tools and resources essential for mastering image SEO. It includes platforms for image compression, AI-driven creation, and educational content. It features DALL-E for AI-generated images and ChatGPT for AI-assisted content and SEO strategies.

Image Compression Tools

- Compressor.io is an efficient tool for compressing images without losing quality, supporting formats like JPEG, PNG, and GIF.
    - URL: https://compressor.io
- Tinify specializes in compressing PNG and JPEG files to significantly reduce file size while maintaining image quality.
    - URL: https://tinypng.com

AI-Driven Image Creation Tools

- DALL-E by OpenAI is an AI system capable of creating unique images from textual descriptions, useful for generating custom, SEO-optimized images. At the time of this writing, you need ChatGPT-4 to use this application. It does not come with the free ChatGPT-3.5. You have to upgrade to ChatGPT-4 (in my opinion a great investment).
    - URL: https://chat.openai.com
    - Please visit OpenAI's website (https://openai.com) for the latest information on accessing DALL-E.
- Canva incorporates AI elements to help design professional-looking images, offering templates and tools that align with SEO principles.
    - URL: https://canva.com

AI-Assisted SEO and Content Creation Tools

- ChatGPT by OpenAI is useful for generating SEO-optimized content, including image descriptions, alt texts, and content strategy planning.

    - URL: https://chat.openai.com
    - More information can be found directly on OpenAI's website (https://openai.com).

Visual Search and Analysis Tools

- Google Images provides image search capabilities that help users understand how images are indexed and ranked in search engines.
    - URL: https://images.google.com

Educational Resources and Communities

- Moz offers comprehensive guides and articles on SEO, including specific strategies for optimizing images.
    - URL: https://moz.com/learn/seo
- Search Engine Journal is a leading publication for the latest in SEO, including trends and best practices in image optimization.
    - URL: https://www.searchenginejournal.com

By using these tools and resources, SEO practitioners can improve their image optimization efforts. They can do this by using AI for content and image creation. They can also stay informed on the latest trends and best practices in SEO.

Appendix B

Image SEO Checklist

This appendix provides a checklist to ensure that images are fully optimized for SEO. This tool is designed to help webmasters, SEO specialists, and content creators. It helps them check that they have covered every aspect of image optimization.

Technical Optimization

- Image Compression: Ensure images are compressed to reduce file size without significantly compromising quality.

- File Format Selection: Choose the appropriate file format (JPEG, PNG, WebP, etc.) based on the image content and usage.

- Responsive Design: Implement responsive image techniques so images display correctly on all devices and screen sizes.
- Loading Speed: Check that images do not negatively impact page loading times, using lazy loading if necessary.

SEO Metadata

- Descriptive File Names: Use clear, descriptive file names that include relevant keywords.
- Alt Text: Add concise, descriptive alt text to all images, incorporating target keywords where appropriate.
- Title and Caption: Where applicable, use titles and captions to provide context and enhance keyword relevance.
- Image Sitemap: Include images in your website's sitemap to facilitate better indexing by search engines.

Content Relevance

- Contextual Alignment: Ensure images are contextually relevant to the surrounding content.

- Image Quality: Use high-quality images that are visually appealing and relevant to the content topic.

- Text in Images: Avoid using important text within images; ensure critical content is in HTML format to be easily indexed.

Advanced Techniques

- Structured Data: Implement structured data (schema markup) for images where applicable to enhance visibility in search results.

- Visual Search Optimization: Optimize images for visual search capabilities, considering clarity, uniqueness, and relevance.

- AI Optimization: Utilize AI tools for optimizing image attributes and enhancing overall image SEO performance.

Legal and Ethical Considerations

- Copyright Compliance: Ensure all images are either owned, licensed, or legally sourced, with appropriate credits and attributions.

- Ethical Use: Consider the ethical implications of image content, ensuring it's culturally sensitive and contextually appropriate.

Monitoring and Maintenance

- Regular Audits: Conduct periodic image SEO audits to identify and rectify issues, ensuring ongoing optimization.

- Performance Tracking: Monitor image search performance and user engagement metrics to gauge the effectiveness of SEO strategies.

- Update and Refresh: Regularly update and refresh images to maintain relevance and engagement over time.

This checklist is a complete guide. It ensures that all critical parts of image SEO are considered and optimized. This helps to get better search rankings and engage users with your website's images.

Glossary

This glossary provides definitions and examples for key terms related to image SEO. It will enhance the reader's understanding and use of the concepts in the book.

- **Alt Text (Alternative Text)**: A textual description of an image used by search engines and screen readers for accessibility. Example: For an image of a sunset over the ocean, the alt text could be "Vibrant sunset over the calm ocean."

- **Compression**: The process of reducing the size of an image file to decrease load time on a web page without significantly affecting its quality. Example: Turning a 5MB JPEG image into a 500KB file through compression tools like Adobe Photoshop or Compressor.io.

- **EXIF Data (Exchangeable Image File Format)**: Information stored within an image file, including details like camera settings, date, time, and location where the photo was taken. Example: A digital photo may have EXIF data showing it was taken with a

Nikon D850 at 1/200 sec, f/1.8, ISO 64, and in New York City.

- **File Format**: The structure and type of an image file, which impacts its use, quality, and size. Common formats include JPEG, PNG, GIF, and WebP. Example: A logo might be saved in PNG format to preserve transparency, while a photograph might be in JPEG format for better compression.

- **Image Alt Attributes**: HTML attributes used to specify alternate text for an image if it cannot be displayed. Example: In HTML, an image tag might look like <img src="dog.jpg" alt="Golden retriever playing fetch">, where the alt attribute describes the image.

- **Image SEO (Search Engine Optimization)**: The practice of optimizing images so they are easily indexed and ranked by search engines, enhancing the visibility of the website they are on. Example: Optimizing an image by using a descriptive file name, reducing the file size, and adding relevant alt text and captions.

- **Image Sitemap**: An XML sitemap exclusively for images, which helps search engines find and index images more effectively. Example: An e-commerce site creates an image sitemap listing all product images with detailed captions and metadata to improve indexing.

- **Lossless Compression**: Reducing an image's file size without losing any quality, ideal for images requiring high detail and clarity. Example: A PNG image of a technical diagram is compressed from 1MB to 600KB with no loss in detail.

- **Lossy Compression**: A method of reducing an image's file size by slightly lowering its quality, which can significantly decrease the file size. Example: A JPEG photograph of a landscape is reduced in file size from 3MB to 500KB, with some loss of subtle details.

- **Metadata**: Additional information about an image, such as the title, description, and keywords, which helps search engines understand and categorize the content. Example: A photo has a title like "Tabby Cat Sleeping." The description says, "A cute tabby cat sleeps curled up in a sunny spot." It has keywords like "cat," "tabby," "sleeping," and "pet."

- **Responsive Images**: Images that adapt in size and resolution based on the screen size and resolution of the viewing device. Example: A website's hero image that displays at full resolution on a desktop but scales down for optimal viewing on smartphones.

- **SEO-Friendly File Names**: The practice of naming image files in a way that reflects the content of the image and is optimized for search engines. Example: Naming an image file "fresh-organic-apples.jpg" rather than "IMG_12345.jpg" for an image depicting fresh apples.

- **Thumbnail Optimization**: The process of optimizing thumbnail images to be clear and descriptive, often used in search results and on social media. Example: Creating a thumbnail for a YouTube video that clearly shows what the video is about, encouraging users to click and view.

- **Visual Search**: A technology that allows users to search using images instead of words, using image recognition to find related content. Example: Using Google Lens to take a picture of a plant to search for information about its species and care instructions.

These terms and examples provide a foundation for understanding image SEO. They help make optimization strategies more effective and improve search engine performance.

Thank You for Reading

We are finishing our exploration of image SEO. I want to thank you for your commitment to learning about this critical part of digital marketing. Your engagement and willingness to dig deep into image optimization details are what make the journey worthwhile.

Next Steps

If this book has enriched your understanding and given you actionable tips to improve your image SEO, I would greatly appreciate your feedback. Your review is more than a comment. It's a beacon for others navigating the complex waters of SEO. It's a compass that guides the resource's improvement.

Your insights are very valuable. They help not only fellow readers, who benefit from your experience, but also me. I rely on them as I work to improve and expand the content in future editions. Here's how your review can make a difference:

- Illuminate the Path for Others: Help potential readers discern the book's value and applicability to their SEO challenges.
- Foster Community and Growth: Contribute to the broader discourse on image SEO, encouraging innovation and shared learning.
- Shape Future Editions: Provide constructive feedback that can inform enhancements, ensuring the book remains a pertinent resource in the evolving field of SEO.

To leave a review, please visit the website where you bought the book. You can also visit any major book review platform. Share your journey. Explain the impact of the strategies you've used. Also, describe your experience with the book.

For more discussions and professional connection, I invite you to connect with me on LinkedIn. Let's keep talking and shape the future of image SEO together.

Thank you once again for your time and dedication. Here's to your success in mastering image SEO. May it bring you more visibility in the digital world!

Warm regards,

Matthew Lane

https://www.linkedin.com/in/mlane9/